Where COVID Came From

T0161782

Nicholas Wade

WHERE COVID CAME FROM

New York · London

The Covid-19 pandemic has disrupted lives the world over for more than a year. Its death toll will soon reach three million people. Yet the origin of the pandemic remains uncertain: the political agendas of governments and scientists have generated thick clouds of obfuscation, which the mainstream press seems helpless to dispel.

In what follows I will sort through the available scientific facts, which hold many clues as to what happened, and provide readers with the evidence to make their own judgments. I will then try to assess the complex issue of blame, which starts with, but extends far beyond, the government of China.

By the end of this essay, you may have learned a lot about the molecular biology of viruses. I will try to keep this process as painless as possible. But the science cannot be avoided because for now, and probably for a long time hence, it offers the only sure thread through the maze.

The virus that caused the pandemic is known officially as SARS-CoV-2, but it can be called SARS2 for short. As many people know, there are

two main theories about its origin. One is that
SARS2 jumped naturally from wildlife to people.
The other is that the virus was under study in a lab,
from which it escaped. It matters a great deal
which is the case if we hope to prevent a second
such occurrence.

I'll describe the two theories, explain why each
is plausible, and then ask which provides the better explanation of the available facts. It's important
to note that so far there is *no direct evidence* for
either theory. Each depends on a set of reasonable
conjectures but so far lacks proof. So I have only
clues, not conclusions, to offer. But those clues
point in a specific direction. And having inferred
that direction, I'm going to delineate some of the
strands in this tangled skein of disaster.

A tale of two theories

After the pandemic first broke out in December 2019, Chinese authorities reported that many cases had occurred in the wet market – a place selling wild animals for meat – in Wuhan. This reminded experts of the SARS1 epidemic of 2002, in which a bat virus had spread first to civets, an animal sold in wet markets, and then from civets to people. A similar bat virus, known as MERS, caused a second outbreak in 2012. This time the intermediary host animal was camels.

The decoding of the SARS2 virus's genome showed it belonged to a viral family known as beta-coronaviruses, to which the SARS1 and MERS viruses also belong. The relationship supported the idea that, like them, it was a natural virus that had managed to jump from bats, via another animal host, to people. The wet market connection, the only other point of similarity with the SARS1 and MERS epidemics, was soon broken: Chinese researchers found earlier cases in Wuhan with no link to the wet market. But that seemed not to matter when so much further evidence in support of natural emergence was expected shortly.

Wuhan, however, is home of the Wuhan Institute of Virology, a leading world center for research on coronaviruses. So the possibility that the SARS2 virus had escaped from the lab could not be ruled out. Two reasonable scenarios of origin were on the table.

From early on, public and media perceptions were shaped in favor of the natural emergence scenario by strong statements from two scientific groups. These statements were not at first examined as critically as they should have been.

"We stand together to strongly condemn conspiracy theories suggesting that COVID-19 does not have a natural origin," a group of virologists and others wrote in *The Lancet* on February 19, 2020, when it was really far too soon for anyone to be sure what had happened. Scientists "overwhelmingly conclude that this coronavirus originated in wildlife," they said, calling on readers to stand with Chinese colleagues on the frontline of fighting the disease.[1]

Contrary to the letter writers' assertion, the idea that the virus might have escaped from a lab invoked accident, not conspiracy. It surely needed to be explored, not rejected out of hand. A defining mark of good scientists is that they go to great pains to distinguish between what they know and what they don't know. By this criterion, the signatories of the *Lancet* letter were behaving as poor scientists: they were assuring the public of facts they could not know for sure were true.

It later turned out that the *Lancet* letter had been organized and drafted by Peter Daszak, president of the EcoHealth Alliance of New York.[2] Dr. Daszak's organization funded coronavirus research at the Wuhan Institute of Virology. If the SARS2 virus had indeed escaped from research he funded, Dr. Daszak would be potentially at fault. This acute conflict of interest was not declared to *The Lancet*'s readers. To the contrary, the letter concluded, "We declare no competing interests."

Virologists like Dr. Daszak had much at stake in the assigning of blame for the pandemic. For twenty years, mostly beneath the public's attention, they had been playing a dangerous game. In their laboratories they routinely created viruses more dangerous than those that exist in nature. They argued they could do so safely, and that by getting ahead of nature they could predict and prevent natural "spillovers," the cross-over of viruses from an animal host to people. If SARS2 had indeed escaped from such a laboratory experiment, a savage blowback could be expected, and the storm of public indignation would affect virologists everywhere, not just in China. "It would shatter the scientific edifice top to bottom," an *MIT Technology Review* editor, Antonio Regalado, said in March 2020.[3]

A second statement which had enormous influence in shaping public attitudes was a letter (in other words, an opinion piece, not a scientific article) published on March 17, 2020, in the journal *Nature Medicine*.[4] Its authors were a group of virologists led by Kristian G. Andersen of the Scripps Research Institute. "Our analyses clearly show that SARS-CoV-2 is not a laboratory construct or a purposefully manipulated virus," the five virologists declared in the second paragraph of their letter.

Unfortunately, this was another case of poor science, in the sense defined above. Viruses can be manipulated in ways that leave no defining marks. Dr. Andersen and his colleagues were assuring their readers of something they could not know.

The discussion part of their letter begins, "It is improbable that SARS-CoV-2 emerged through laboratory manipulation of a related SARS-CoV-like coronavirus." But wait, didn't they open by saying the virus had "clearly" not been manipulated? The authors' degree of certainty seemed to slip several notches when it came to laying out their reasoning.

The reason for the slippage is clear: the authors offer two technical reasons why manipulation is unlikely, but neither is in any way conclusive. Their blanket assertion that the virus could not have been manipulated, even though grounded in nothing but two inconclusive speculations, convinced the world's press that SARS2 could not have escaped from a lab. A technical critique of the Andersen letter takes it down in harsher words.[5]

Science is supposedly a self-correcting community of experts who constantly check each other's work. So why didn't other virologists point out that the Andersen letter's argument was full of absurdly large holes? Perhaps because in today's universities speech can be very costly. Careers can be destroyed for stepping out of line. Any virologist who challenges the community's declared view risks having his next grant application turned down by the panel of fellow virologists that advises the government grant distribution agency.

Really, the Daszak and Andersen letters were political, not scientific, statements, yet they were amazingly effective. Articles in the mainstream press repeatedly stated that a consensus of experts had ruled lab escape out of the question or extremely unlikely. Their authors relied for the most part on the Daszak and Andersen letters, failing to understand the yawning gaps in their arguments. Mainstream newspapers all have science journalists on their staff, as do the major networks, and these specialist reporters are supposed to be able to question scientists and check their assertions. But the Daszak and Andersen assertions went largely unchallenged.

Doubts about natural emergence

Natural emergence was the media's preferred theory until around February 2021 and the visit by a World Health Organization commission to China. The commission's composition and access were heavily controlled by the Chinese authorities. Its members, who included the ubiquitous Dr. Daszak, kept asserting before, during, and after their visit that lab escape was extremely unlikely. But this was not quite the propaganda victory the Chinese authorities may have been hoping for. What became clear was that the Chinese had no evidence to offer the commission in support of the natural emergence theory.

This was surprising because both the SARS1 and MERS viruses had left copious traces in the environment. The intermediary host species of SARS1 was identified within four months of the epidemic's outbreak, and the host of MERS within nine months. Yet some fifteen months after the SARS2 pandemic began, and a presumably intensive search, Chinese researchers had failed to find either the original bat population, or the intermediate species to which SARS2 might have jumped, or any serological evidence that any Chinese population, including that of Wuhan, had ever been exposed to the virus prior to December 2019. Natural emergence remained a conjecture which, however plausible to begin with, had gained not a shred of supporting evidence in over a year.

And as long as that remains the case, it's logical to pay serious attention to the alternative conjecture, that SARS2 escaped from a lab.

Why would anyone want to create a novel virus capable of causing a pandemic? Ever since virologists gained the tools for manipulating a virus's genes, they have argued they could get ahead of a potential pandemic by exploring how close a given animal virus might be to making the jump to humans. And that justified lab experiments in enhancing the ability of dangerous animal viruses to infect people, virologists asserted.

With this rationale, they have recreated the 1918 flu virus, shown how the almost extinct polio virus can be synthesized from its known RNA sequence, and introduced a smallpox gene into a related virus.

These enhancements of viral capabilities are known blandly as gain-of-function experiments. With coronaviruses, there was particular interest in the spike proteins, which jut out all around the spherical surface of the virus and pretty much determine which species of animal it will target. In 2000, Dutch researchers, for instance, earned the gratitude of rodents everywhere by genetically engineering the spike protein of a mouse corona-virus so that it would attack only cats.[6]

Virologists started studying bat coronaviruses in earnest after these turned out to be the source of both the SARS1 and MERS epidemics. In particular, researchers wanted to understand what changes needed to occur in a bat virus's spike pro-teins before it could infect people.

Researchers at the Wuhan Institute of Virology, led by China's leading expert on bat viruses, Dr. Shi Zheng-li or "Bat Lady," mounted frequent expeditions to the bat-infested caves of Yunnan in southern China and collected around a hundred different bat coronaviruses.

Dr. Shi then teamed up with Ralph S. Baric, an eminent coronavirus researcher at the University of North Carolina. Their work focused on enhanc-ing the ability of bat viruses to attack humans so as to "examine the emergence potential (that is, the potential to infect humans) of circulating bat CoVs [coronaviruses]."[7] In pursuit of this aim, in

November 2015 they created a novel virus by taking the backbone of the SARS1 virus and replacing its spike protein with one from a bat virus (known as SHC014-CoV). This manufactured virus was able to infect the cells of the human airway, at least when tested against a lab culture of such cells.

The potential danger of the novel entity concerned many observers. "If the virus escaped, nobody could predict the trajectory," said Simon Wain-Hobson, a virologist at the Pasteur Institute in Paris.[8]

Dr. Baric and Dr. Shi also recognized the potential risk of this kind of experiment, but they argued that the risks should be weighed against "the potential to prepare for and mitigate future outbreaks."

That statement was made in 2015. From the hindsight of 2021, one can say that the value of gain-of-function studies in preventing the SARS2 epidemic was zero. The risk was catastrophic, if indeed the SARS2 virus was generated in a gain-of-function experiment.

Inside the Wuhan Institute of Virology

Dr. Baric had developed, and taught Dr. Shi, a general method for engineering bat coronaviruses to attack other species. The specific targets were human cells grown in cultures and humanized mice. These laboratory mice, a cheap and ethical stand-in for human subjects, are genetically engineered to carry the human version of a protein called ACE2 that studs the surface of cells that line the airways.

Dr. Shi returned to her lab at the Wuhan Institute of Virology and resumed the work she had started on genetically engineering coronaviruses to attack human cells.

How can we be so sure?

Because, by a strange twist in the story, her work was funded by the National Institute of Allergy and Infectious Diseases (NIAID), a part of the U.S. National Institutes of Health (NIH). And grant proposals that funded her work, which are a matter of public record, describe what she planned to do with the money.

The grants were assigned to the prime contractor, Dr. Daszak of the EcoHealth Alliance, who subcontracted them to Dr. Shi. Abstracts of the grants, which are publicly available, indicate that Dr. Shi planned to insert spike genes from some viruses into the backbone of others, creating a series of chimeric viruses. These chimeric viruses would then be tested for their ability to attack human cell cultures and humanized mice. And this information would help predict the likelihood of "spillover," the jump of a coronavirus from bats to people.

The methodical approach was designed to find the best combination of coronavirus backbone and spike protein for infecting human cells. The approach could have generated SARS2-like viruses, and indeed may have created the SARS2 virus itself with the right combination of virus backbone and spike protein.

It cannot yet be stated that Dr. Shi did or did not generate SARS2 in her lab because her records have been sealed, but it seems she was on the right track to have done so. "It is clear that the Wuhan Institute of Virology was systematically constructing novel chimeric coronaviruses and was assessing their ability to infect human cells and human-ACE2-expressing mice," says Richard H. Ebright, a molecular biologist at Rutgers University and leading expert on biosafety.

"It is also clear," Dr. Ebright says, "that, depending on the constant genomic contexts chosen for analysis, this work could have produced SARS-CoV-2 or a proximal progenitor of SARS-CoV-2." "Genomic context" refers to the particular viral backbone used as the testbed for the spike protein.

Even if the grant required the work plan described above, how can we be sure that the plan was in fact carried out? For that we can rely on the word of Dr. Daszak, who has been much protesting for the last fifteen months that lab escape was a ludicrous conspiracy theory invented by China-bashers.[9]

On December 9, 2019, before the outbreak of the pandemic became generally known, Dr. Daszak gave an interview in which he talked in glowing terms of how researchers at the Wuhan Institute of Virology had been reprogramming the spike protein and generating chimeric coronaviruses capable of infecting humanized mice: "And we have now found, you know, after six or seven years of doing this, over one hundred new SARS-related coronaviruses, very close to SARS," Dr. Daszak says around twenty-eight minutes into the interview.[10] "Some of them get into human cells in the lab, some of them can cause SARS disease in humanized mice models and are untreatable with therapeutic monoclonals and you can't vaccinate against them with a vaccine. So, these are a clear and present danger."

The interviewer asks: "You say these are diverse coronaviruses and you can't vaccinate against them, and no anti-virals – so what do we do?"

Daszak: "Well I think … coronaviruses – you can manipulate them in the lab pretty easily. Spike protein drives a lot of what happen with coronavirus, in zoonotic risk. So you can get the sequence, you can build the protein, and we work a lot with Ralph Baric at UNC to do this. Insert into the backbone of another virus and do some work in the lab. So you can get more predictive when you find a sequence. You've got this diversity. Now the logical progression for vaccines is, if you are going to develop a vaccine for SARS, people are going to use pandemic SARS, but let's insert some of these other things and get a better vaccine." The insertions he referred to perhaps included an element called the furin cleavage site, discussed below, which greatly increases viral infectivity for human cells.

In disjointed style, Dr. Daszak is referring to the fact that once you have generated a novel coronavirus that can attack human cells, you can take the spike protein and make it the basis for a vaccine.

One can only imagine Dr. Daszak's reaction when he heard of the outbreak of the epidemic in Wuhan a few days later. He would have known better than anyone the Wuhan Institute's goal of making bat coronaviruses infectious to humans, as well as the weaknesses in the institute's defense against their own researchers becoming infected.

But instead of providing public health authorities with the plentiful information at his disposal, he immediately launched a public relations campaign to persuade the world that the epidemic couldn't possibly have been caused by one of the institute's souped-up viruses. "The idea that this virus escaped from a lab is just pure baloney. It's simply not true," he declared in an April 2020 interview.[11]

The lab escape scenario for the origin of the SARS2 virus, as should by now be evident, is not mere hand-waving in the direction of the Wuhan Institute of Virology. It is a detailed proposal, based on the specific research program being pursued by Dr. Shi.

The safety arrangements
at the Wuhan Institute of Virology

Dr. Daszak was possibly unaware of, or perhaps
he knew all too well, the long history of viruses
escaping from even the best-run laboratories.[12]
The smallpox virus escaped three times from labs
in England in the 1960s and '70s, causing eighty
cases and three deaths. Dangerous viruses have
leaked out of labs almost every year since. In more
recent times, the SARS1 virus has proved a true
escape artist, leaking from laboratories in Singa-
pore, Taiwan, and no less than four times from the
Chinese National Institute of Virology in Beijing.

One reason for SARS1 being so hard to handle
is that there were no vaccines available to protect
laboratory workers. As Dr. Daszak mentioned in
his December 19 interview quoted above, the
Wuhan researchers too had been unable to
develop vaccines against the coronaviruses they
had designed to infect human cells. They would
have been as defenseless against the SARS2 virus,
if it were generated in their lab, as their Beijing
colleagues were against SARS1.

A second reason for the severe danger of novel

coronaviruses has to do with the required levels of lab safety. There are four degrees of safety, designated BSL1 to BSL4, with BSL4 being the most restrictive and designed for deadly pathogens like the Ebola virus.

Before 2020, the rules followed by virologists in China and elsewhere required that experiments with the SARS1 and MERS viruses be conducted in BSL3 conditions. But all other bat coronaviruses could be studied in BSL2, the next level down. BSL2 requires taking fairly minimal safety precautions, such as wearing lab coats and gloves, not sucking up liquids in a pipette, doing lab work under a hood, and putting up biohazard warning signs. Yet a gain-of-function experiment conducted in BSL2 might produce an agent more infectious than either SARS1 or MERS. And if it did, then lab workers would stand a high chance of infection.

Much of Dr. Shi's work on gain-of-function in coronaviruses was performed at the BSL2 safety level, as is stated in her publications and other documents. She has said in an interview with *Science* magazine that "The coronavirus research in our laboratory is conducted in BSL2 or BSL3 laboratories."[13]

"It is clear that some or all of this work was being performed using a biosafety standard –

biosafety level 2, the biosafety level of a standard US dentist's office – that would pose an unacceptably high risk of infection of laboratory staff upon contact with a virus having the transmission properties of SARS-CoV-2," says Dr. Ebright.

Concern about safety conditions at the Wuhan lab was not, it seems, misplaced. According to a fact sheet issued by the State Department on January 15, 2021, "The U.S. government has reason to believe that several researchers inside the WIV became sick in autumn 2019, before the first identified case of the outbreak, with symptoms consistent with both COVID-19 and common seasonal illnesses."[14]

David Asher, a fellow of the Hudson Institute and former consultant to the State Department, provided more detail about the incident at a seminar.[15] Knowledge of the incident came from a mix of public information and "some high-end information collected by our intelligence community," he said. Three people working at a BSL3 lab at the institute fell sick within a week of each other with severe symptoms that required hospitalization. This was "the first known cluster that we're aware of, of victims of what we believe to be COVID-19." Influenza could not completely be ruled out but seemed unlikely in the circumstances, he said.

Comparing the rival scenarios of SARS2 origin

The evidence above adds up to a serious case that the SARS2 virus could have been created in a lab, from which it then escaped. But the case, however substantial, falls short of proof. Proof would consist of evidence from the Wuhan Institute of Virology, or related labs in Wuhan, that SARS2 or a predecessor virus was under development there. For lack of access to such records, another approach is to take certain salient facts about the SARS2 virus and ask how well each is explained by the two rival scenarios of origin, those of natural emergence and lab escape. Here are three tests of the two hypotheses.

1. The place of origin

Start with geography. The two closest known relatives of the SARS2 virus were collected from bats living in caves in Yunnan, a province of southern China. If the SARS2 virus had first infected people living around the Yunnan caves, that would strongly support the idea that the virus had spilled over to people naturally. But this isn't what happened. The pandemic broke out 1,500 kilometers away, in Wuhan.

What if the bat viruses infected some intermediate host first? You would need a longstanding population of bats in frequent proximity with an intermediate host, which in turn must often cross paths with people. All these exchanges of virus must take place somewhere outside Wuhan, a busy metropolis which so far as is known is not a natural habitat of *Rhinolophus* bat colonies. The infected person (or animal) carrying this highly transmissible virus must have traveled to Wuhan without infecting anyone else. No one in his or her family got sick. If the person jumped on a train to Wuhan, no fellow passengers fell ill.

It's a stretch, in other words, to get the pandemic to break out naturally outside Wuhan and then, without leaving any trace, to make its first appearance there.

For the lab escape scenario, a Wuhan origin for the virus is a no-brainer. Wuhan is home to China's leading center of coronavirus research where, as noted above, researchers were genetically engineering bat coronaviruses to attack human cells. They were doing so under the minimal safety conditions of a BSL2 lab. If a virus with the unexpected infectiousness of SARS2 had been generated there, its escape would be no surprise.

2. *Natural history and evolution*

The initial location of the pandemic is a small part of a larger problem, that of its natural history. Viruses don't just make one-time jumps from one species to another. The coronavirus spike protein, adapted to attack bat cells, needs repeated jumps to another species, most of which fail, before it gains a lucky mutation. Mutation – a change in one of its RNA units – causes a different amino acid unit to be incorporated into its spike protein and makes the spike protein better able to attack the cells of some other species.

Through several more such mutation-driven adjustments, the virus adapts to its new host – say, some animal with which bats are in frequent contact. The whole process then resumes as the virus moves from this intermediate host to people.

In the case of SARS1, researchers have documented the successive changes in its spike protein as the virus evolved step by step into a dangerous pathogen. After it had gotten from bats into civets, there were six further changes in its spike protein before it became a mild pathogen in people. After a further fourteen changes, the virus was much better adapted to humans, and with a further four the epidemic took off.[16]

But when you look for the fingerprints of a similar transition in SARS2, a strange surprise awaits. The virus has changed hardly at all, at least until recently. From its very first appearance, it was well adapted to human cells. Researchers led by Alina Chan of the Broad Institute compared SARS2 with late stage SARS1, which by then was well adapted to human cells, and found that the two viruses were similarly well adapted. "By the time SARS-CoV-2 was first detected in late 2019, it was already pre-adapted to human transmission to an extent similar to late epidemic SARS-CoV," they wrote.[17]

Even those who think lab origin unlikely agree that SARS2 genomes are remarkably uniform. Dr. Baric writes that "early strains identified in Wuhan, China, showed limited genetic diversity, which suggests that the virus may have been introduced from a single source."[18]

A single source would of course be compatible with lab escape, less so with the massive variation and selection that is evolution's hallmark way of doing business.

The uniform structure of SARS2 genomes

gives no hint of any passage through an intermediate animal host, and no such host has been identified in nature.

Proponents of natural emergence suggest that SARS2 incubated in a yet-to-be found human population before gaining its special properties. Or that it jumped to a host animal outside China.

All these conjectures are possible, but strained. Proponents of the lab leak theory have a simpler explanation. They say that SARS2 was adapted to human cells from the start because it was grown in humanized mice or in lab cultures of human cells, just as described in Dr. Daszak's grant proposal. Its genome shows little diversity because the hallmark of lab cultures is uniformity.

Proponents of laboratory escape propose sardonically that of course the SARS2 virus infected an intermediary host species before spreading to people, and that they have identified it – a humanized mouse from the Wuhan Institute of Virology.

3. *The furin cleavage site*

Within the SARS2 virus's anatomy, there is a small region of its spike protein, called the "furin cleavage site," which is coded for by twelve units of its 30,000-unit genome.

A virus usually acquires inserts like this by accidentally exchanging genomic units with another virus when both invade the same cell. But no other known virus in SARS2's group (Sarbecoviruses) has this twelve-unit insert.

Proponents of natural emergence argue that the virus could have acquired the insert from human cells after it had jumped to people. Maybe, but no one has yet found the human population in which the virus might have evolved this way. The insert also contains entities known as arginine codons, which are common in humans but not in coronaviruses like SARS2.

On the lab escape scenario, the insert is easy to explain. "Since 1992 the virology community has known that the one sure way to make a virus deadlier is to give it a furin cleavage site," writes Dr. Steven Quay, a biotech entrepreneur interested in the origins of SARS2.[19] At least eleven such experiments have been published, including one by Dr. Shi.

"When I first saw the furin cleavage site in the viral sequence, with its arginine codons, I said to my wife it was the smoking gun for the origin of the virus," said David Baltimore, an eminent virologist and former president of the California Institute of Technology. "These features make a powerful challenge to the idea of a natural origin for SARS2," he said.

Where we are so far

Neither the natural emergence nor the lab escape hypothesis can yet be ruled out. There is still no direct evidence for either. So no definitive conclusion can be reached.

That said, the available evidence leans more strongly in one direction than the other. Readers will form their own opinion. But it seems to me that proponents of lab escape can explain all the available facts about SARS2 considerably more easily than can those who favor natural emergence.

Researchers at the Wuhan Institute of Virology were doing gain-of-function experiments designed to make coronaviruses infect human cells and humanized mice. This is exactly the kind of experiment from which a SARS2-like virus could have emerged. The researchers were not vaccinated against the viruses under study, and they were working in the minimal safety conditions of a BSL2 laboratory. So escape of a virus would not be at all surprising. In all of China, the pandemic broke out on the doorstep of the Wuhan institute. The virus was already well adapted to humans, as expected for a virus grown in humanized mice. It possessed an unusual enhancement, a furin cleavage site, which is not possessed by any other known SARS-related beta-coronavirus, and this site included a double arginine codon also unknown among beta-coronaviruses. What more evidence could you want, aside from the presently unobtainable lab records documenting SARS2's creation?

Proponents of natural emergence have a rather harder story to tell. The plausibility of their case rests on a single surmise, the expected parallel between the emergence of SARS2 and that of SARS1 and MERS. But none of the evidence expected in support of such a parallel history has yet emerged. No one has found the bat population that was the source of SARS2, if indeed it ever infected bats. No intermediate host has presented itself, despite an intensive search by Chinese authorities that included the testing of eighty thousand animals. There is no evidence of the virus making multiple independent jumps from its intermediate host to people, as both the SARS1 and MERS viruses did. There is no evidence from hospital surveillance records of the epidemic gathering strength in the population as the virus evolved. There is no explanation of why a natural epidemic should have broken out in Wuhan and nowhere else. There is no good explanation of how the virus acquired its furin cleavage site, which no other SARS-related beta-coronavirus possesses, nor why the site is composed of human-preferred codons. The natural emergence theory battles a bristling array of implausibilities.

The records of the Wuhan Institute of Virology certainly hold much relevant information. But Chinese authorities seem unlikely to release them given the substantial chance that they incriminate the regime in the creation of the pandemic. Absent the efforts of some courageous Chinese whistle-blower, we may already have at hand just about all of the relevant information we are likely to get for a while.

So it's worth trying to assess responsibility for the pandemic, at least in a provisional way, because the paramount goal remains to prevent another one. Even those who aren't persuaded that lab escape is the more likely origin of the SARS2 virus may see reason for concern about the present state of regulation governing gain-of-function research. Here are the players who seem most likely to deserve blame.

1. Chinese virologists

First and foremost, Chinese virologists are to blame for performing gain-of-function experiments in mostly BSL2-level safety conditions that were far too lax to contain a virus of unexpected infectiousness like SARS2. If the virus did indeed escape from their lab, they deserve the world's censure for a foreseeable accident that has already caused the deaths of three million people.

True, Dr. Shi was trained by French virologists, worked closely with American virologists, and was following international rules for the containment of coronaviruses. But she could and should have made her own assessment of the risks she was running. She and her colleagues bear the responsibility for their actions.

I have been using the Wuhan Institute of Virology as a shorthand for all virological activities in Wuhan. It's possible that SARS2 was generated in some other Wuhan lab, perhaps in an attempt to make a vaccine that worked against all coronaviruses. But until the role of other Chinese virologists is clarified, Dr. Shi is the public face of Chinese work on coronaviruses, and provisionally she and her colleagues will stand first in line for opprobrium.

2. Chinese authorities

China's central authorities did not generate SARS2, but they sure did their utmost to conceal the nature of the tragedy and China's responsibility for it. They suppressed all records at the Wuhan Institute of Virology and closed down its virus databases. They released a trickle of information, much of which may have been outright false or designed to misdirect and mislead. They did their best to manipulate the WHO's inquiry into the virus's origins, and they led the commission's members on a fruitless run-around. So far they have proved far more interested in deflecting blame than in taking the steps necessary to prevent a second pandemic.

3. The worldwide community of virologists

Virologists around the world are a loose-knit professional community. They write articles in the same journals. They attend the same conferences. They have common interests in seeking funds from governments and in not being overburdened with safety regulations.

Virologists knew better than anyone the dangers of gain-of-function research. But the power to create new viruses, and the research funding obtainable by doing so, was too tempting. They pushed ahead with gain-of-function experiments. American virologists lobbied against the moratorium imposed on U.S. federal funding for gain-of-function research in 2014, and it was raised in 2017.

The benefits of the research in preventing future epidemics have so far been nil, the risks vast. If research on the SARS1 and MERS viruses could only be done at the BSL3 safety level, it was surely illogical to allow any work with novel coronaviruses at the lesser level of BSL2. Whether or not SARS2 escaped from a lab, virologists around the world have been playing with fire.

Their behavior has long alarmed other biologists. In 2014, scientists calling themselves the Cambridge Working Group urged caution on creating new viruses. In prescient words, they specified the risk of creating a SARS2-like virus. "Accident risks with newly created 'potential pandemic pathogens' raise grave new concerns," they wrote.[20] "Laboratory creation of highly transmissible, novel strains of dangerous viruses, especially but not limited to influenza, poses substantially increased risks. An accidental infection in such a setting could trigger outbreaks that would be difficult or impossible to control."

When molecular biologists discovered a technique for moving genes from one organism to another, they held a public conference at Asilomar in 1975 to discuss the possible risks. Despite much internal opposition, they drew up a list of stringent safety measures that could be relaxed in future – and duly were – when the possible hazards had been better assessed.

When the CRISPR technique for editing genes was invented, biologists convened a joint report by the U.S., U.K., and Chinese national academies of science to urge restraint on making heritable changes to the human genome. Biologists who invented gene drives have also been open about the dangers of their work and have sought to involve the public.

You might think the SARS2 pandemic would spur virologists to re-evaluate the benefits of gain-of-function research, even to engage the public in their deliberations. But no. Many virologists deride lab escape as a conspiracy theory and others say nothing. They have barricaded themselves behind a Chinese wall of silence which so far is working well to allay, or at least postpone, journalists' curiosity and the public's wrath. Professions that cannot regulate themselves deserve to get regulated by others, and this would seem to be the future that virologists are choosing for themselves.

4. The United States' role in funding the Wuhan Institute of Virology

From June 2014 to May 2019 Dr. Daszak's EcoHealth Alliance had a grant from the National Institute of Allergy and Infectious Diseases (NIAID), part of the National Institutes of Health (NIH), to support research with coronaviruses at the Wuhan Institute of Virology. Whether or not SARS2 is the product of that research, it seems a questionable policy to farm out high-risk research to unsafe foreign labs using minimal safety precautions. And if the SARS2 virus did indeed escape from the Wuhan institute, then the NIH could find itself in the terrible position of having funded a disastrous experiment that led to death of more than three million worldwide, including more than half a million of its own citizens.

At present writing, however, it seems clear that the NIH did not intend its funds to be used for gain-of-function. There was a moratorium on funding such research that lasted from 2014 to 2017 and the agency would presumably have been in violation of the law had its funds gone for any such purpose. Both Dr. Anthony Fauci, director of the NIAID, and Dr. Francis Collins, director of the NIH, have stated that they funded no gain-of-function research at the Wuhan institute.

But from Dr. Daszak's description in his December 2019 interview of swapping around spike genes from one coronavirus to another, it's reasonable to assume that his sub-grantee, Dr. Shi, was creating viruses more powerful than those that exist in nature. This raises the possibility, yet to be determined, that Dr. Shi was using her research funds for purposes not authorized by the NIH.

In conclusion

If the case that SARS2 originated in a lab is so substantial, why isn't this more widely known? As may now be obvious, there are many people who have reason not to talk about it. The list is led, of course, by the Chinese authorities. But virologists in the United States and Europe also have no great interest in igniting a public debate about the gain-of-function experiments that their community has been pursuing for years.

Nor have other scientists stepped forward to raise the issue. Government research funds are distributed on the advice of committees of scientific experts drawn from universities. Anyone who rocks the boat by raising awkward political issues runs the risk that their grant will not be renewed, and that their research career will be ended.

The United States government shares a strange common interest with the Chinese authorities: neither is keen on drawing attention to the fact that Dr. Shi's coronavirus work was funded by the U.S. National Institutes of Health. One can imagine the behind-the-scenes conversation in which the Chinese government says "If this research was so dangerous, why did you fund it, and on our territory too?" To which the U.S. side might reply, "Looks like it was you who let it escape. But do we really need to have this discussion in public?"

To these serried walls of silence must be added that of the mainstream media. To my knowledge, no major newspaper or television network has yet provided readers with an in-depth news story of the lab escape scenario, such as the one you have just read, although some have run brief editorials or opinion pieces. One might think that any plausible origin of a virus that has killed three million people would merit a serious investigation. Or that the wisdom of continuing gain-of-function research, regardless of the virus's origin, would be worth probing. Or that the NIH's funding of what sounds like gain-of-function research during a moratorium on such funding would bear investigation. What accounts for the media's apparent lack of curiosity?

The virologists' omertà is one reason. Science reporters, unlike political reporters, have little innate skepticism of their sources' motives; most see their role largely as purveying the wisdom of scientists to the unwashed masses. So when their sources won't help, these journalists are at a loss.

Another reason, perhaps, is the migration of much of the media toward the left of the political spectrum. Because President Trump said the virus had escaped from a Wuhan lab, editors gave the idea little credence. They joined the virologists in regarding lab escape as a dismissible conspiracy theory. During the Trump Administration, they had no trouble in rejecting the position of the intelligence services that lab escape could not be ruled out. But when Avril Haines, President Biden's director of National Intelligence, said the same thing, she too was largely ignored. This is not to argue that editors should have endorsed the lab escape scenario, merely that they should have explored the possibility fully and fairly.

People round the world who have been pretty much confined to their homes for the last year might like a better answer than their media are giving them. Perhaps one will emerge in time. After all, the more months that pass without the natural emergence theory gaining a shred of supporting evidence, the less plausible it may seem. Perhaps the international community of virologists will come to be seen as a false and self-interested guide. The common-sense perception that a pandemic breaking out in Wuhan might have something to do with a Wuhan lab cooking up novel viruses of maximal danger in unsafe conditions could eventually displace the ideological insistence that whatever Trump said can't be true.

And then let the reckoning begin.

Acknowledgments

The first person to take a serious look at the origins of the SARS2 virus was Yuri Deigin, a biotech entrepreneur in Russia and Canada. In a long and brilliant essay, he dissected the molecular biology of the SARS2 virus and raised, without endorsing, the possibility that it had been manipulated.[21] The essay, published on April 22, 2020, provided a roadmap for anyone seeking to understand the virus's origins.

In Deigin's wake have followed several other skeptics of the virologists' orthodoxy. Nikolai Petrovsky calculated how tightly the SARS2 virus binds to the ACE2 receptors of various species and found to his surprise that it seemed optimized for the human receptor, leading him to infer the virus might have been generated in a laboratory. Alina Chan published a paper showing that SARS2 from its first appearance was very well adapted to human cells.[22] [23]

One of the very few establishment scientists to have questioned the virologists' absolute rejection of lab escape is Richard Ebright, who has long warned against the dangers of gain-of-function research. Another is David A. Relman of Stanford University. "Even though strong opinions abound, none of these scenarios can be confidently ruled in or ruled out with currently available facts," he wrote.[24] Kudos too to Robert Redfield, former director of the Centers for Disease Control and Prevention, who told CNN on March 26, 2021 that the "most likely" cause of the epidemic was "from a laboratory," because he doubted that a bat virus could become an extreme human pathogen overnight, without taking time to evolve, as seemed to be the case with SARS2.[25]

Steven Quay, a physician-researcher, has applied statistical and bioinformatic tools to ingenious explorations of the virus's origin, showing, for instance, how the hospitals receiving the early patients are clustered along Line 2 of the Wuhan subway system, which connects the Institute of Virology at one end with the international airport at the other, the perfect conveyor belt for distributing the virus from lab to globe.[26]

In June 2020, Milton Leitenberg published an early survey of the evidence favoring lab escape from gain-of-function research at the Wuhan Institute of Virology.[27]

Many others have contributed significant

pieces of the puzzle. "Truth is the daughter," said Francis Bacon, "not of authority but time." The efforts of people such as those named above are what makes it so.

Notes

1   Charles Calisher et. al., "Statement in support of the scientists, public health professionals, and medical professionals of China combatting COVID-19," *The Lancet*, Vol. 395, Issue 10226, https://doi.org/10.1016/S0140-6736(20)30418-9.

2   Sainath Suryanarayanan, "EcoHealth Alliance orchestrated key scientists' statement on 'natural origin' of SARS-CoV-2," *U.S. Right to Know*, November 18, 2020, https://usrtk.org/biohazards-blog/ecohealth-alliance-orchestrated-key-scientists-statement-on-natural-origin-of-sars-cov-2/.

3   Antonio Regalado (@antonioregalado), "While genetic engineering is unlikely," Twitter, April 27, 2020, https://twitter.com/antonioregalado/status/1254916969712803840?lang=en.

4   Kristian G. Andersen et. al., "The proximal origin of SARS-CoV-2," *Nature Medicine*, Vol. 26, 450-452 (March 2020), https://doi.org/10.1038/s41591-020-0820-9.

5   Harvard to the Big House, "China owns *Nature* magazine's ass: Debunking 'The proximal origin of SARS-CoV-2' claiming COVID-19 definitely wasn't from a lab," March 19, 2020, https://harvardtothebighouse.com/2020/03/19/china-owns-nature-magazines-ass-debunking-the-proximal-origin-of-sars-cov-2-claiming-covid-19-wasnt-from-a-lab/.

6   Lili Kuo et. al., "Retargeting of coronavirus by substitution of the spike glycoprotein ectodomain: crossing the host cell species barrier," *Journal of Virology*, Vol. 74(3), 1393-1406 (February 2000), doi: 10.1128/jvi.74.3.1393-1406.2000.

7   Vineet D. Menachery et. al., "A SARS-like cluster of circulating bat coronaviruses shows potential for human emergence," *Nature Medicine*, Vol. 21, 1508-1513 (November 2015), https://doi.org/10.1038/nm.3985.

8   Declan Butler, "Engineered bat virus stirs debate over risky research," *Nature*, November 12, 2015, https://www.nature.com/news/engineered-bat-virus-stirs-debate-over-risky-research-1.18787.

9   Peter Daszak, "Ignore the conspiracy theories: scientists know Covid-19 wasn't created in a lab," *Guardian*, June 9, 2020, https://www.theguardian.com/commentisfree/2020/jun/09/conspiracies-covid-19-lab-false-pandemic.

10 Vincent Racaniello, This Week in Virology, "TWiV 615: Peter Daszak of EcoHealth Alliance," YouTube video, May 19, 2020, https://www.youtube.com/watch?v=IdYDL_RK--w.

11 Amy Goodman and Nermeen Shaikh, "'Pure Baloney': Zoologist debunks Trump's Covid-19 origin theory, explains animal-human transmission," interview transcript, *Democracy Now!* April 16, 2020, https://www.democracynow.org/2020/4/16/peter_daszak_coronavirus.

12 Martin Furmanski, "Laboratory Escapes and 'Self-Fulfilling Prophecy' Epidemics," February 17, 2014, https://armscontrolcenter.org/wp-content/uploads/2016/02/Escaped-Viruses-final-2-17-14-copy.pdf.

13 Reply to *Science* magazine, Dr. Shi Zhengli interview, https://www.sciencemag.org/sites/default/files/Shi%20Zhengli%20Q%26A.pdf.

14 U.S. Department of State, Office of the Spokesperson, "Fact Sheet: Activity at the Wuhan Institute of Virology," January 15, 2021, https://2017-2021.state.gov/fact-sheet-activity-at-the-wuhan-institute-of-virology/index.html.

15 David Asher and Miles Yu, "Transcript: The origins of Covid-19: Policy implications and lessons from the future," seminar transcript, Hudson Institute, March 12, 2021, https://www.hudson.org/research/16762-transcript-the-origins-of-covid-19-policy-implications-and-lessons-for-the-future.

16 Biao Kan et. al., "Molecular evolution analysis and geographic investigation of severe acute respiratory syndrome coronavirus-like virus in palm civets at an animal market and on farms," *Journal of Virology*, Vol. 79 (18), 11892-11900 (August 2005), doi: 10.1128/JVI.79.18.11892-11900.2005.

17 Shing Hei Zhan, Benjamin E. Deverman, and Yujia Alina Chan, "SARS-CoV-2 is well adapted for humans. What does this mean for re-emergence?" bioRXiv.org, May 2020, https://www.biorxiv.org/content/10.1101/2020.05.01.073262v1.

18 Ralph S. Baric, "Emergence of a highly fit SARS-CoV-2 variant," *The New England Journal of Medicine*, 383: 2684-2686, December 31, 2020, https://www.nejm.org/doi/10.1056/NEJMcibr2032888.

19 Steven Carl Quay, "An introduction to a Bayesian analysis of the laboratory origin of SARS-CoV-2,", .mp4 download, January 29, 2021, https://zenodo.org/record/4477212#.YKvwoOspAhe.

20 "Cambridge Working Group Consensus Statement on the Creation of Potential Pandemic Pathogens (PPPs)," *The Cambridge Working Group*, July 14, 2014.

21 Yuri Deigin, "Lab Made? SARS-CoV-2 Genealogy Through the Lens of Gain-of-Function Research," *Yuri Deigin* (blog), April 22, 2020, https://yurideigin.medium.com/lab-made-cov2-genealogy-through-the-lens-of-gain-of-function-research-f96dd7413748.

22 Piplani et. al, "In silico comparison of SARS-CoV-2 spike protein-ACE2 binding affinities across species and implications for viral origin," (joint senior academic paper), https://arxiv.org/ftp/arxiv/papers/2005/2005.06199.pdf.

23 Yujia Alina Chan, Benjamin E. Deverman, and Shing Hei Zhan, "SARS-CoV-2 is well adapted for humans. What does this mean for re-emergence?" bioRXiv.org, May 2, 2020, https://doi.org/10.1101/2020.05.01.073262.

24 David A. Relman, "Opinion: To stop the next pandemic, we need to unravel the origins of COVID-19," Proceedings of the National Academy of Sciences of the United States, November 3, 2020, https://doi.org/10.1073/pnas.2021133117.

25 CNN's *New Day*, hosted by John Berman et al. Interview of CDC director Robert Redford by Dr. Sanjay Gupta.

26 Steven Carl Quay, "Where did the 2019 coronavirus pandemic begin and how did it spread? The people's liberation army hospital in Wuhan China and line 2 of the Wuhan metro system are compelling answers," October 28, 2020, https://zenodo.org/record/4119263#.YKvyCOspAhd.

27 Milton Leitenberg, "Did the SARS-CoV-2 virus arise from a bat coronavirus research program in a Chinese laboratory? Very possibly," *Bulletin of the Atomic Scientists*, June 4, 2020, https://thebulletin.org/2020/06/did-the-sars-cov-2-virus-arise-from-a-bat-coronavirus-research-program-in-a-chinese-laboratory-very-possibly/.

A NOTE ON THE TYPE

WHERE COVID CAME FROM has been set in Scala Sans, designed in the 1990s by Martin Majoor as a companion to his Scala serif fonts of the 1990s. A distinctly contemporary face, Scala marries the proportions of old-style types to the monoline strokes of geometric sans-serifs. The especially pleasing harmony between roman and italic reveals the designer's attention to every detail of the faces. Although its ancestry can be traced to no single family of types, Scala succeeds in a wide range of uses by virtue of its clean drawing, its vertical emphasis, and its personable, somewhat casual letterforms. ◇ The display type is Berthold's Akzidenz Grotesk Light Condensed, a twentieth-century addition to a highly regarded family of types whose origins date to 1896.

DESIGN & COMPOSITION BY
CARL W. SCARBROUGH

First American edition published in 2021 by Encounter Books, an activity of Encounter for Culture and Education, Inc., a nonprofit, tax-exempt corporation.
Encounter Books website address: www.encounterbooks.com

Manufactured in the United States and printed on acid-free paper. The paper used in this publication meets the minimum requirements of ANSI/NISO Z39.48–1992 (R 1997) (*Permanence of Paper*).

FIRST AMERICAN EDITION

LIBRARY OF CONGRESS
CATALOGING-IN-PUBLICATION DATA
IS AVAILABLE